The Fantastic Book of Biology Jokes: For Everyone, Not Just Biologists

UNCONVENTIONAL PUBLISHING

www.unconventionalpublishing.com.au

Shane Van

ISBN – 978-0-6452206-8-1 HARDCOVER

DISCLAIMER: THIS BOOK IS A WORK OF FICTION AND IS NOT TO BE TAKEN SERIOUSLY; THERE IS NOTHING IN THIS BOOK THAT SHOULD BE TAKEN AS FACT, ESPECIALLY SCIENTIFIC FACT.

TABLE OF CONTENTS

Preface

Unconventional Publishing proudly present a book that will make you laugh and think of science and biology differently. Nothing in this book should be taken as fact and is only here to put a smile on your face; no part of this book is meant to offend. If it does in any way offend you, you can always put it down and move on to something else.

This book is the second in a series of books and follows on from *The Fantastic Book of Chemistry Jokes.* This series roasts all sciences and everyday professions. The series will cover professions from chemists to physicists, doctors to plumbers.

The Wild Forest

You must be an Auxin, because you're causing my stem to elongate.

What did the stamen say to the pistil?
You are on point with your style.

Why can't plants escape jail?
Their cells are surrounded by walls.

Where does a dendrochronologist keep his notes?
In a tree-ring binder.

Why didn't the dendrochronologist ever get married?
He only dated trees.

Why do plants like to photosynthesis?
So, they can have a light snack.

What kind of instruments do botanist's play?
Woodwinds.

There was a man who decided to open a plant nursery. He was working hard in everything he did and put a lot of money into the business. He wasn't making any profit no matter what he tried. Eventually, the creditors came calling and started looking through his books to see what they could salvage so he could repay the loan.

The reappraiser was walking through the nursery inspecting the property when he turns to the owner and asks, "Why did you want to open up a nursery?"

"Well, I've always wanted to own a business, be my own boss," he replies

"Did you ever study plants or have an interest in them?"

"Not really, I just saw a cheap business for sale and decided to have a go."

The reappraiser then has a look at the stock room, looks around, nods, then writes something in his book. He then enters the office and starts looking at the expense book.

"Ok, I figured out why your business failed. You know the plants? You haven't botany."

What place is world-renowned to produce the best botanists?

The University of Flora-da.

Boll weevils are detrimental to crops, they can ruin a whole year's work in a matter of days. The farmers get inundated with small weevils and large weevils. If you had the choice of 2 weevils which one, would you choose?
The small weevils, it is the lesser of two weevils.

What did the botanist say when he saw 2 peanut bushes mating?
Wow, that's indehiscent.

Botanists pick up lines;

- I will never leaf you.
- You make my heart skip a beet.
- Wood you be mine?

Two botanists are sitting at the bar, one looks really upset. The other one asks, "Hey, what stomata with you?"

Why is grass so dangerous?
Because it is full of blades

A couple of botanists were in a long-term relationship, their hobby was gardening their own rare and exotic plants. This hobby was an expensive one as a lot of plants need specialised soils and constant monitoring. It cost so much and took up so much time it stopped them from having a family. The wife turned to her husband one day and says, "We can keep the garden orchid."

What is the scariest plant?
BAM-BOO!!!!

What is another word for photosynthesis?
Sun-stenance.

Why are botanists always lost in the woods?
It's hard to see the florist for the trees

I know a botanist that works with nuts,
He lives in macadamia.
I have noticed someone coming into my garden lately and has been adding soil to it. It seems the plot thickens.

There was an Irish botanist who was trying to crossbreed poison ivy and a four-leaf clover.
He only wanted a rash of good luck.

Sexism is everywhere, even in botany. Even to this day, there is so much stigma attached to female plants.

I once met a time-travelling botanist, he kept going back to the fuchsia.

Why did the botanist get an award?
He was outstanding in his field.

What do you call a botanist who also moonlights as an escort?
Whorticulturist.

Why can't a plant ever join the dark side?
They can't make food without light.
I once met a botanist who faked his degree, his thesis was based on a plastic Christmas in bought in a garage sale.

A botanist runs into a bank wearing a balaclava.
"Everyone on the floor, I have a pistil!"

There was once a stand-up comic who was also a botanist. He was telling joke after joke for 10 minutes straight. Everyone was just sitting there looking bored, it was a bit embarrassing. Well, everyone except for this one guy but he must have been a plant.

If you can't name the parts of a flower in a biology exam, you can just always guess the anther.

My biology teacher asked me, what my favourite vegetable was. Apparently, Steven Hawking was an inappropriate answer.

What type of plant can you grow in your hand?
Palm trees.
After a hurricane what trees bounce back the quickest?
Rubber trees.

What is a tree's least favourite month?
Sep-timber.

What do trees wear to swimming pools?
Trunks.

Why did the leaf visit the doctor?
Because it was feeling green.

A Toads Stalk

Mycologists are the noblest of the biologists, they are obsessed with morels.

What does a mushroom say when he needs help?
Help me I am in truffle.

What did a mushroom say to the toadstool after they were licking each other out?
Umami.

Why are fungi not allowed on a bus?
They take up too mushroom.

I like to put my mushrooms in coke when I cook them, you can taste my-cola-gist.

A mycologist tried to become a stand-up comedian; nobody laughed at any of his jokes. Later in the evening he came back and asked a few of the audience members what he did wrong?
Apparently, all his jokes were in spore taste.
My Uncle use to hate eating mushrooms, but now that he's dead they are beginning to grow on him.

What do you call a mushroom that creates music?
A decomposer

Today I posted a video of my wife scratching her pussy. It went fungal.

I once got a fungal infection; it really grew on me.

A priest and a nun were discussing some elements of the Old Testament.
"What would you do if you came across the burning bush?" he asked
"I would put some fungal cream on it immediately!"

Why does Mario hang out with Toad instead of Luigi?
Because he's a fungi.

What did the eukaryote become when he realised, he wasn't a fungi, plant, or animal?
A protist- er.

What did the mushroom get arrested for?
He was a stalker.

Why are mushroom hunters so selfish?

It’s called mycology, not your-cology.

What do you call a mushroom that keeps the boys happy?
Fungal.

What do you call a mushroom with a 12-inch stalk?
Fungi to date.

Why are mushrooms always invited to parties?
Because they are fungi’s.

Why are toadstools not invited to parties?
Because they are toxic.

A biologist tried to become a comedian and made-up thousands of jokes about mushrooms. He was a fungi.

Two funguses are cuddling one night, it starts getting a bit hot in the room. The first fungi says, “Why don’t you move over?”
“I can’t there isn’t too mush-room.”

What kind of car do mushrooms drive?
A spore’s car.

Did you hear about the 2 sourdough loaves that fell in love?

They raised some dough, put a bun in the oven and grew mould together.

What did Mike Tyson say when he saw mould on his bananas?

That's growth.

A friend of mine went to their daughter's dollhouse, ripped off the roof to search for mould.

I thought that was a little spore-attic.

George goes to the urinal and finds a gentleman standing there with no arms. The guy turns to George and says, "I am so sorry, but can you please help me?"

Feeling sorry for the poor chap he says, "Sure, what can I do."

"Um well, can you unzip me?"

So, George unzips him.

"Um, could you also pull it out and hold it for me as well?"

So, George feeling sorry for the guy, does it. As he is doing it, he notices that the guy's penis is all red and covered in weeping puss and lumps with a horrible cheese smell. George swallows down and reluctantly just holds it for the guy as he urinates.

"Thank you, do you mind shaking it and putting it back in?"
So, George does and then hurries up and washes his hands. He can't stop wondering.
"Sorry, but, what is wrong with your penis?" he asked.
"No Idea but I ain't touching it," the man says as he pulls his arms out of his sleeves.

My new fridge detects mould, it has a spoiler alert.

Bleach kills everything, germs, mould, mildew, black t-shirts, and blue jeans but not white clothes. God damn white privilege.

Where do sores stay when they go for holidays?
In a mould-tel.

Why do biologists have a successful marriage? They like to grow mold together.

A mother was cleaning out her son's bag, when she sees a sandwich at the bottom. This sandwich had been sitting there for months. It was mostly black and grey with lots of furs.
"What the hell is this?" she yells at her son,
"Are you trying to breed penicillin?"

On Safari

During a discussion on lizard behaviour, a zoologist pointed out that when there is cold weather moves in, the iguana's freeze and then fall out of the trees.
A student gets up and runs out, yelling, "Iguana see this."

How do zoologists tell crocodiles and alligators apart?
It depends if it sees you in a while crocodile or later alligator.

How do you turn on a zoologist?
Send them your dik-dik pics.

You remind me of a pair of Parus Major, you've got great tits.

Let me tell you about my new girlfriend, she's a vegetarian. You've probably never heard of herbivore.

This man had just broken up with his girlfriend and was feeling lonely when he wakes up one morning hearing a faint buzz. Shaking his head it stops, but slowly over time, it starts getting louder and louder. It started a little in the morning and then it would be random throughout the day until it was almost all the time during the day and night. He couldn't sleep he was feeling nauseous all the time and he finally went to the doctor.

The doctor asks him about his symptoms. They were nausea, dizziness headaches and hallucinations. He also mentions he has been extremely tired but couldn't sleep because of the noise. He stammers and explains that something weird is also happening, his ears are leaking wax and it's really sweet.

The doctor looks at him with seriously concerned eyes and asks, "Where do you live?"

"Hiko." says the patient

The doctor then goes with his stethoscope and checks the man's chest. All he hears is buzzing like a beehive.

"Ok I am sorry but have something to tell you and it's not pleasant," says the doctor. "There is this secret army base up in the hills behind Hiko. Area 51. They were experimenting with genetically modified bees. These bees were super-intelligent, and their job was infecting soldiers and turning these soldiers into super soldiers. The bees didn't want any part of this and escaped. Unfortunately, they do still need human bodies to survive. They try their best to keep the host alive but eventually they will kill the host. There is nothing we can do; you will need to get your business in order. I am sorry."

The man leaves the doctors in disbelief, he buys a bottle of whiskey and goes home. He sits on the veranda, has a sip of whiskey, neat and lights a cigarette. Just staring into space in disbelief when his phone rings. It is his ex-girlfriend and she wanted to get back together.

He says, "Bee still my beating heart!!!"

What is the most fertile place in South America?

Spermatogonia

There was a recent news article in Africa, on how a group of zoologists were putting on shaggy costumes and covering themselves in antelope urine. It was just so they could get closer to beasts and observe them. Turns out it was just fake gnus.

How do you identify a bald eagle?
His feathers have a bad combover.

Eggs are now on the endangered species list, it is because of excessive poaching

Where do hippos' study?
In the hippocampus.

Two zoologists are having a bit of a discussion amphibians. One asks the other,
"Did you read the book about frogs?"
"Yes, I found it absolutely ribbiting!"

What did the French zoologist hope to see on his date?
Ze-bra.

A zoologist was studying monkeys and how they could be classified as entering the stone age because they started using simple tools like stone axes. He then wanted to test what would happen if he gave them some advanced technology. He ended up arming them with rifles and pistols. Soon the monkeys were divided into groups and fighting over territory. He named what they were doing 'gorilla warfare'.

Three moles form a straight line and start digging a hole, the first mole stops digging and starts sniffing the air.
"I smell chocolate."
The second mole then stops and sniffs the air
"I smell toast."
The last mole then stops and sniffs the air
"I smell molasses."

Fun fact, a pack of piranhas can pick apart a school kid and leave nothing but bone in 30 seconds.
....and that's why I lost my job at the zoo.

What do you get when you cross an elephant with a giraffe?
A Nobel Prize in biology

A zoologist was called out to investigate why hundreds of bison were turning up dead near all the highways. Upon investigation, it was determined that all the bison were actually killed by trucks and only trucks. With further investigation, they found out that it was because of the crows. While they could say, "Caw" they couldn't say "Trucks."

Hippos are the most dangerous animals in Africa, in fact, they kill more people than any other animal on the continent. This gives zoologists a major problem when estimating the size of their backs. They call it finding hippotenuse.

When I was 6 years old, the priest used to take me by the hand into the back of the woods. He did this with several children for 2 years until he disappeared. Turns out he became a zoologist, he just loved teaching kids about animals.

What did the zoologist and a botanist name their child?
Tiger Woods.

A frog decides to find out more about his past and gets a genealogy test.
He discovered he was a tad-pole-ish.

What replaced tapeworm?
The CD worm.

A drunk zoologist walks into a bar with a crocodile under his arm. With everyone looking at him, he staggers up to the bar, throws the crocodile on it and orders a drink.
"Get that thing off of my bar!" yells the bartender.
"Who? Tiny? He's harmless." slurs the zoologist, "I'll prove it."

The toucan had to become the top of its environmental niche. It did have stiff competition but eventually, it did win over the toucannot.

He then smacks the crocodile on the head, and it opens its mouth. He then pulls out his penis and puts it in the crocodile's mouth. With one hand on his hip, his penis in the crocodile's mouth and with the other hand sculling a beer, he then slams the beer on the table, pulls out his penis, slaps the crocodile on the head again and it closes its mouth.
"Would anyone else want a turn?" he says proudly.
This old man at the back of the bar gets up.
"I will but you don't need to hit me on the head."

A frog goes to a fortune-teller.
"I can see a bright future for you, you will be the centre of attention, a young woman will want to know everything about you"
"Oh, will this happen at a party?"
"No, in a biology class."

My zoologist friend told me that he is working on a new project. It involves reclassifying all the giraffes into species based on their neck types. I told him that was a tall order.
2 elephants were discussing what they are doing in their lives now.
The first elephant says, "I am researching my ancestors."
"That must be a mammoth task." says the second elephant.

A zoologist does his morning rounds when he notices the kangaroo out of his cage just pounding through the enclosures. He herds the kangaroo back into his cage and then extends the fence height was from 10 to 15ft high. The next morning the zoologist is doing the rounds again, sees the kangaroo in amongst the elephants, pounding on their backs. Angrily, the zoologist builds the fence up to 20 ft and herds the kangaroo back in.

As he is building the fence a giraffe leans across and asks the kangaroo.
"How long do you think this will go on for?"
"Well, he'll probably give up around the 45ft mark and then send me back home or he'll learn to lock the gate."
Two zoologists are in the woods, radio-tracking a bear. A bear manages to sneak up on them and comes crashing through the woods. The zoologists both climb up the nearest tree as fast as they could. The bear runs to the tree and tries to climb, roaring and leaping at their feet. It then starts grabbing the tree, biting the trunk and trying to push it over. One zoologist takes off his hiking boots and throws them at the bear. It hits it straight in the head. He then pulls joggers out of his bag and puts them on.
"What the hell are you doing?" asks, the second biologist.
"Well, when this tree breaks, we are going to have to hit the ground running."
"You know we don't have a chance to outrun a bear."
"No, no. I just need to outrun you."

Under the Sea

Two surfers who hadn't seen each other in ages were meeting up on the beach.
"Hey Brah, I haven't seen you in ages." the first surfer says.
"Yeah Brah, I wanted to study marine biology so I could learn more about the ocean." says the second surfer.
"Really Brah?"
"Yeah, Brah but it was too much maths, there was a lot of algae-Brah."

What compounds washes up on beaches? Nucleotides.

Why did the scuba diver fail biology?
He was below C level.

How does a marine biologist say goodbye?
Sea you later.

When I started to study marine biology, I felt I finally found my porpoise in life.
What is the difference between a marine biologist and a dog?
The dog wags a tail, the other tags a whale.

A new scientific study on marine biology went viral, it also caused many marine biologists to start huge arguments. The study revealed that the widespread coral bleaching on The Great Barrier was actually caused by migrating crows.
The main criticism was that the coralation does not imply cawsation.

Why do great innovators make great marine biologists?
They can observe all the e-fish-they-see.

Biologists have discovered an evolutionary trait in many sea mammals. When they need rest, one side of the brain falls asleep, and one side stays awake. They then swaps sides. This is a survival mechanism. Biologists have also noted this trait in bureaucrats the whole world over.

How many ecologists does it take to change a lightbulb?
Zero, they would never replace a broken light bulb, think of the carbon footprint.

A research team of marine biologists, accidentally catch a dolphin in their nets. They are about to release when they see it has feet!! What a strange mutation. They study it, look at it, discuss it and argue about it.

One of them then yells out, "Why don't we just cut off its feet?"
"Well, that would defeat the porpoise!"

A marine biologist found a group of killer whales that met up once a month and all sang together and made music. They called the event an orca-stra.

What is a marine biologist's favourite instrument?
The tambo-marine.

How do use to cut a wave in half?
A sea-saw.
A group of marine biologists were watching a documentary about sharks. There was one shark that would just swim up, bite the toes of swimmers, then swim off again. This behaviour had them puzzled, then they realised it...
The shark was lack toes intolerant.

A marine biologist studying great white sharks was asked about their diet.
"Well, it consists of marine biologists studying great white sharks."

My brother was studying marine biology, he has been so lazy. He has just been coasting all year.

What hair style does the ocean have?
Wavy.

I took marine biology, it was disappointing, Le Pen wasn't discussed even once!

What do you get if you cross whale DNA with human DNA?
Kicked out of SeaWorld.

The Planet's Cycles

Why do ecologists love lepers?
Because they are biodegradable.

The ecologist's wife left him. All he was doing was trying to save water by showering with his undergrads.

If you ride a bike twice does that make it recycling?

What is the most influential culture?
Agriculture

A small ATM has 2 air conditioners, four fluorescent lights and is turned on 24 hours a day and then it asks me if I would like a receipt to save the environment?

How does President Biden plan to fight the record high temperatures?
By changing from Fahrenheit to Celsius.

What is a cross between red and yellow, about 70 years old, has caused enormous environmental damage, thousands of illnesses and is a great embarrassment to a country?

Agent Orange

Where do hobbits live?
In a hobbitat.

I was with a friend the other day who was using a bunch of plastic straws. I told him that he shouldn't be doing that.
"YEAH, YEAH, I know. Save the turtles and shit. I get it!" He replied
"It's not that, it's just a really weird way to eat spaghetti."

The ecology teacher was showing slides of different birds and asking students to name them.
"Does anyone know the name of this particular bird?"
"Sir it's a Willow Ptarmigan"
"Ok, now class did you know you can never hear one use the bathroom? Does anyone know why?"
The class just looks at each other and no one answer.
"It's because the p is silent."

Halloween is just around the corner, people should be thinking more about the environment, those plastic skeletons cause mass amounts of pollution. Locally sourced, all-natural skeletons are much better.

A biology student was talking to his friends and telling them how he was a natural ecologist and suited to his environment.
His friend laughed and asked him to go on.
"Well," the student continues; "remember that time we tried weed, and I was so stoned? I went home and my parents were waiting for me. I thought I was going to get busted but they started talking to me. I kinda phased in and out but turns out I am adapted."

Hear about the ecologist who didn't have a carbon footprint?
It's because he drove everywhere.
Clean coal is like wearing a condom with holes in it. At least the intention is there.

Who is more environmentally friendly, Apple or IBM?
Well, Apple has a trash can and IBM has a recycle bin.

Why did the sun go to school?
To become brighter.

Turns out Stalin was an environmentalist. Killing millions of your own people will help the environment.

What is an ecologist's least favourite dish?
Baked Alaska.

Petrol prices are hiking through the roof, in California, the women can't even run over their husbands in their Mercedes'.

Where do Snowmen party?
At the ball.
An ecologist decided that to help keep the carbon footprint down decided he would carpool with his colleagues. He picked up Simon and then George. When George was getting into the car, he was a little bit hesitant, but he hopped in. Unfortunately, on this trip there were road works and a detour onto the highway, which made George even more anxious, he was visually shaking.
When they saw the tunnel, he started screaming, "STOP THE CAR, STOP THE CAR!"
Turns out George was a sufferer of carpool tunnel syndrome.

What do you call an ecologist who cuts down a tree?
Treeson!

An ecologist had twins; she named the first child carbon. She named the second child copy.

Why is smoking good for the environment?
Because it kills humans.

I am tired of people telling me to keep my lights off to save energy, last time I did that I nearly ran over some pedestrians.

There once was this Australian woman, an environmentalist, an avid protestor and anti-hunter who bought a large plot of land with a protected old-growth forest on it. She decided to climb the biggest, oldest gum tree so she could see her property.
As she climbed, this koala came out and attacked her, she leaned back, then slid off the tree, getting splinters all in and around her vagina while she fell.
Painfully she went to the ER and explained her situation and what had happened to the nurse. The nurse then grabbed the doctor and frustratedly the environmentalist repeated her story. The doctor replied with a little mutter of 'I see' then wandered off.
A few hours later, the woman is still sitting there and now getting even angrier when the doctor comes back in.

"Where the BLOODY HELL have you been!" she demanded.
"You see, this is more of a complicated issue than you understand, I had to ring Parks and Wildlife services, the Environmental Protection Authority and the Council's Land Management Planner. I had to get permits to remove old growth timbers from a recreational area close to a waste disposal outlet. Unfortunately, I couldn't get all the permits and can't proceed."

Two ecologists were having a chat.
"So, what is your view on renewable energy?" asks the first.
"I don't know about you, but I am a big fan." says the second.

Where do ecologist's like to go for holidays?
Mad-at-gas-car.

There is a new TV show that purely discusses renewable energy, it's called the solar panel.

What do stand-up comedians and ecologists have in common?
They are both good at recycling.

A biologist phones his wife from the office.
"I am so sorry honey, I have to go out into the field this week. Doctor Jones has fallen over and broken his ankle. I am going to have to leave this afternoon, I will come past at lunch and pick up my suitcase and field kit, can you please pack them for me? I am so sorry about this."
So, he swings home, picks up his suitcase and his field kit, kisses his wife goodbye and sets off for a week.
When he returns he sees his wife.
"Hi babes, thank you so much for packing for me but you didn't pack my Pj's."
"Yes, I did, they were in your field kit."

Have you heard about the power plant that was bad all year?
Santa gave him coal for Christmas.

An ecology student had to write a report on wind energy, it was a breeze.

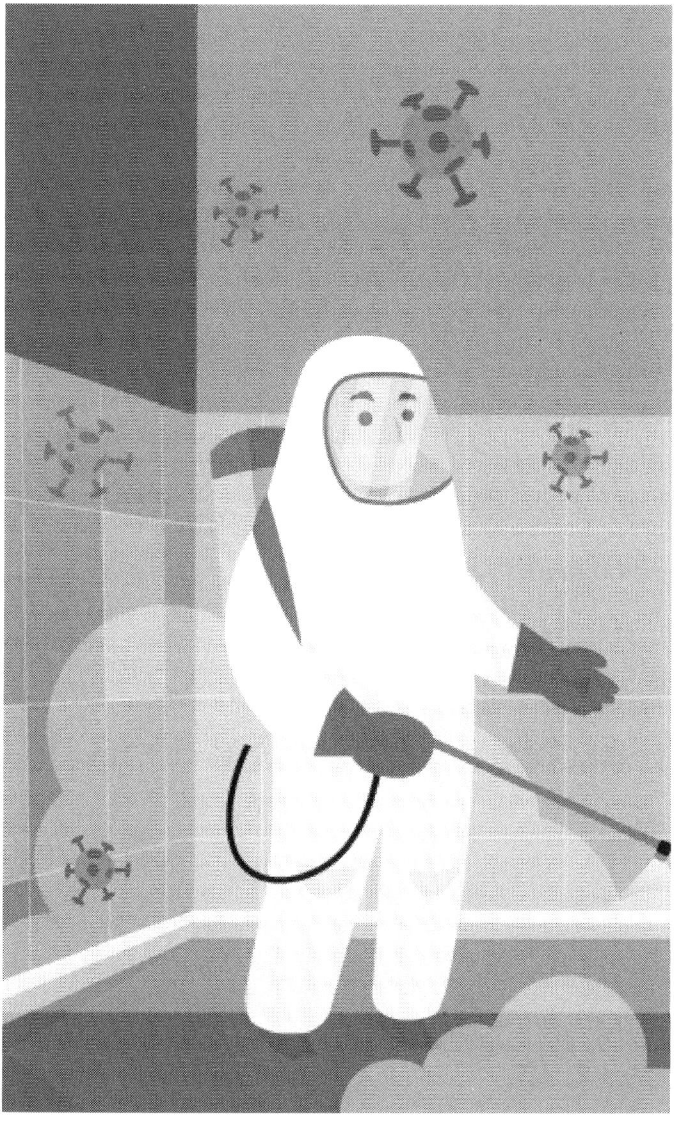

The Tiny World

Why was the biologist constantly overlooked?
He was only a micro-biologist

Say what you want about microbiologists, but they really know how to appreciate the little things in life.

How many microgiologists does it take to change a light bulb?
That depends on the population density.

I think bacteria is just extraordinary, it can be found in the harshest environment and survive extinction events, bacteria are used in cheese making. It just reminds me that where there is life. there is another whey.

A guy and a girl are talking at a bar
"What do you do for a living?" she asks.
"I am a microbiologist, which involves a lot of streaking, probing and mounting, but I do it all with culture and sensitivity"

A few cells go into a bar one night, they have a few drinks, share a few laughs, and leave without causing any trouble. It's because they are cultured cells.

When all the bacteria don't like something, what do they do?
They protist.

What is a paramecium?
Two Italian mice.

Why are microbiologists so smart?
They always have germs of ideas.

I failed my biology exam; I was asked what the most lethal parasite is. Apparently, politicians wasn't the correct answer.

Interesting bit of information…cyanobacteria was the first bacteria that evolved photosynthesis. Back eons ago the earth's atmosphere had very little oxygen in the atmosphere. The place was filled with anaerobic bacteria. Over millions of years, these cyanobacteria changed the atmosphere and caused the extinction of other bacteria. It makes me feel a little better since humans aren't the first creatures to change the atmosphere and cause mass extinction.

A doctor of virology is proud of his university accomplishments in fact in his office he had all his degrees framed and on the wall. He had a BSc of Biology, next to his Master of Biology and then the PhD plaque hung between them.
One day his office caught on fire, and he panicked, he rushed inside to get his accolades when he was overcome by smoke and passed out. He then later woke up at the hospital covered in bandages. The nurse came to him. "You're very lucky you know, you only suffered 1st, 2nd and 3rd-degree burns."

Did you hear about the microbiologist that travelled through every continent and spoke 8 languages?
He had many cultures.

Why did the bacteria cross the microscope?
To get to the other slide.

The biggest problem with antibiotics is that no matter how good they are, they are never going viral.

What is the name of a urinary tract infection in the virgin Mary?
Immaculate Infection.

I thought my biologist wife was cheating on me, so I set up a hidden webcam. Sure enough, she was being pounded without a condom. It went viral.

Did you hear about the germ that evolved into a man?
He became a politician.
If you're a germaphobe and you know it
wash your hands. .📢 ♫ Dum Dum

What do you call a progressive person who is also a germaphobe?
A social distance warrior.

Why was the amoeba sad?
His parents had split.

A germ strolls into a bar, the bartender says, "Hey you get out!"
The germ yells back, "I am staph."

Do you want to hear the joke about germs?
Sorry, I don't want to spread it around.

They say money is covered in germs, but I don't think germs could live on the money I make.

The fear of germs is getting out of control, we are now told that we should sing Bohemian Rhapsody twice when we wash our hands.

My father's sister is totally obsessed with killing all the germs in her house.
We call her Auntie-Bacterial.

What do you call a micro-organism that listens to classical music?
Bach-teria.

Why don't ants get sick?
Because they are tiny antiebodies.

What did the infectious disease say when the bartender kicked him out?
"You're not a very good host."

Where did the virus go?
It flu away.

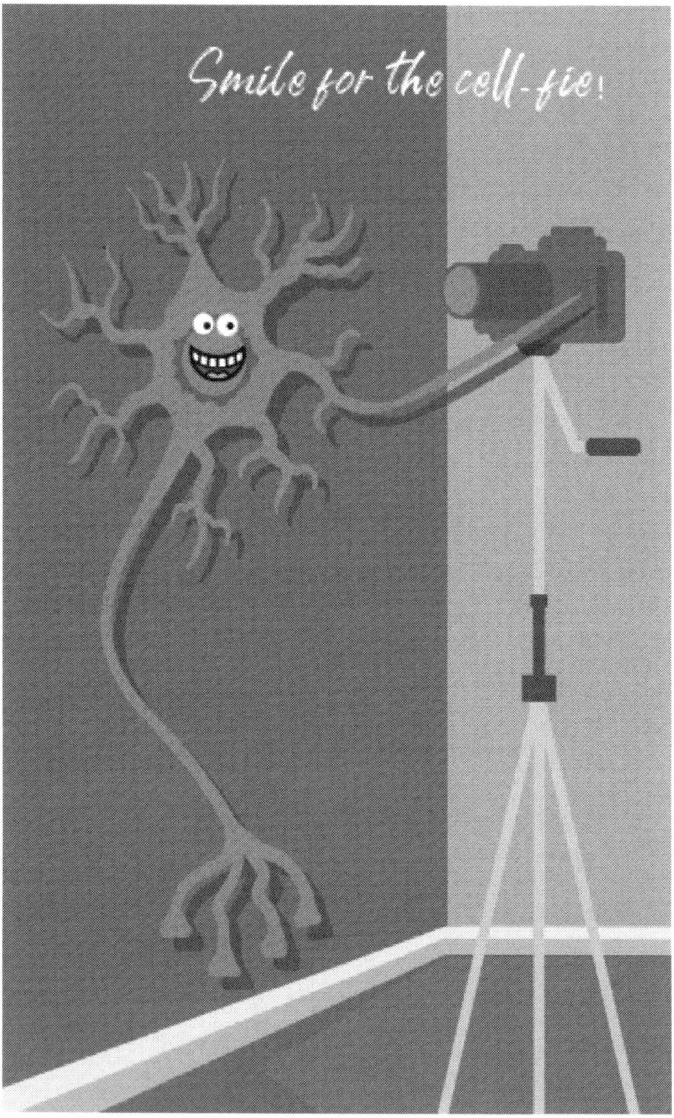
Smile for the cell-fie!

Bad Biology

Life is a sexually transmitted disease.

What is the leader of a gang of biologists called?
The nucleus.

What did the biologist say when asked how she liked her eggs?
She said ovaries-y.

Biology: the only science where divide and multiply mean the same thing.

A man working in a lab, went to grab a glass of water. He accidentally ended up ingesting some alpha-L-glucose but luckily, he didn't suffer any consequences. It turns out he was ambidextrose.

Call me DJ Enzyme, because I break it down

What is a pirate's favourite amino acid? Arrrrrrrrrr----gine.
What did one chromosome say to the other chromosome?
Do I look fat in these genes?

What did the cell say when it ran into the table? Mitosis.

How do you make a hormone?
Don't pay her.

How do you make a hormone?
Kick her in the crotch.

Do you know which part of my body makes a hormone?
My penis.

Why did the woman break up with the biologist?
He was too cell-fish.

What do biologist's wear on casual Fridays? Their genes.

A biologist was working in Australia, she was telling a story about being bitten by 2 spiders.
"The first time I was bitten was back in England."
A colleague piped in, "So did the other one bite you down under?"

What do hipster biologists wear on casual Fridays?

Skinny genes.

What does a biologist wear to impress on a date?
Designer genes.

What do biologists post on Snapchat?
Cell-fies.

What do you call proteins with attitude?
A-mean-old-acid.

I failed my genealogy test, turns out I can't grant one single wish.

What would you use to eat DNA spaghetti?
A replication fork.
What is a conservation biologists' motto?
The only cleavage I want to see should be at a cellular level.

My phone is an organic lifeform. It has cellular activity.

How many geneticits does it take change just 1 light bulb?
23 pairs.

Where do criminal neurons end up?
In a chain ganglion.

What do you get when you drink Pepsi under a full moon?
A peptide.

What is usually found in a cell?
Dumb criminals.

WOAH WOAH, did you just mutate a stop codon?
Because you're not making any sense.

Your mother was that slow it took her 9 months to make a joke.

I have always wondered whether biology is the study of living organisms, or is the study of 2 ology's?

What do prison and biology have in common?
Cell nature.

What is the difference between jail and biology?
In biology, you learn that the mitochondria is the powerhouse of the cell. In jail, you find out Bubba is.

When I failed biology in school, I didn't realise in the end it was going to fail me as well.

Why do women find biology professors irresistible?
They are great at handling multiple organisms.

Your momma is so fat that when you are going down on her, she needs her own mega transect.
What do my socks and cellular division have in common?
Mitosis.

A geneticist makes a startling discovery, he found out how to crossbreed any two organisms. He set up his lab and gets himself an intern to help with his work. He spends a week training the intern and how to work the instruments and teaches them the theory behind what he is doing.
His intern asks him, "Can you really cross any two organisms on the planet?"
"I sure can" he replied, "but the technology is extremely dangerous, especially to us." The geneticist replies.
"Dangerous, how so?"
"The last person that crossed Bill Clinton and Donald Trump ended up hung in a jail cell."

What is the best way biologists communicate with each other?
They use a cellphone.

What do biologists take at Raves?
mDNA.

What do you call the Jewish prophet that passed water through a semi-permeable membrane?
OsMoses.

How does Juliet keep her body in balance?
Romeo-stasis.

What is the fastest way to tell the sex of a chromosome?
Pull down their genes.

What do you call a cow that has genetically evolved?
A moo-tant.

Dick is short for Richard, but that's probably just genetics.

How does a geneticist eat ice cream?
Helix it.

What does a geneticist call going down on two women?
A double helix.

Somebody I once met had a genetic disposition to diarrhoea. It ran in his genes.

If you cloned Henry the IV, would he still be called Henry the IV, Henry the IX, or Henry the IV part 2?

If there were markets for genetic foods, would there be factory seconds?

A guy walks into a greengrocer and looks at the avocados.
"Excuse me sir, are these avocadoes genetically modified?" he asks the cashier
"Why do you ask?" says the cashier
"Yeah, why do you ask?" says the avocado.

A cell undergoing mitosis looks at the biologist and says, 'now I've got your divided attention."

A missionary goes to a tropical island and tries to convert the local population to Christianity. He has been on this island for many months and is now considered one of the locals. One day he is just relaxing in his hut when the chieftain kicks down the door.
“You white bastard! I’ll KILL YOU. You say you are a holy man, but you are a liar and trickster. For that, you will die!!!”
“Why mighty chieftain? What have I done?” pleaded the missionary.
“You bastard, you got my daughter pregnant! That brings shame to the family, and her husband wants to have war with our family. Her son is white!!!” Shouted the chieftain.
“It just happens in nature sometimes, it is a rare thing, but sometimes genetics just misalign, or something else happens, and the colours come out different. Take a look at the pigs out there. They are all pink, except for this dark brown one. It happens.
The chieftain then pats him on the back.
“Ok, ok, I'll keep your secret, and you keep mine.”
Where does a geneticist keep his vegetables?
In the fridges CRISPR draw.

There is a team of geneticists working at KFC.
They are busy trying to make the chicken CRISPR.

12 years after a lady has her first child, she starts looking at her son, he doesn't resemble either her or her husband, in fact, he doesn't even look like any of their other relatives. She orders a DNA test and sure enough, their son is absolutely no blood relation to either of them.
She waits until her husband gets home to talk to him.
"Sweetheart, I have something really serious to discuss with you," she said.
"Why? What's wrong?"
"I had some suspicions and so I did a DNA test for our son. It turns out he isn't related to us."
"Of course, he isn't", the husband says now quite annoyed, he continues, "You don't remember do you?"
"Remember what?" she says.
"When I was wheeling you out the hospital, you were carrying the baby, then he pissed and shit himself and you threw him at me and told me to go change him."

What is a geneticists' favourite shape?
A punnet square.

My grandfather was once told a joke about genetics, my father didn't get it, but I did.

Roses are red

Violets are blue
Roses are also white and pink
Genetics.

A guy comes running into a doctor surgery
"Doctor, doctor." He yells, "My DNA has confused itself."
"AND?" says the doctor.

A geneticist once spliced the DNA of a potato with his genitals. No one likes a dic- tater.

Genetics tells us that you should look like your parents, if not, then it tells you, that you should.

One ejaculation can contain 1547 TB of data, shouldn't that make porn stars geniuses?

A single sperm can contain 37.5MB of data. That means in the average ejaculation there could be 1547.5 TB of data. I don't know about you but that's a lot to swallow.

Did you hear about the baker that was trapped by his genetics?
He was in-bread.

Why is it so hard to solve a red-neck murder?
Because all the DNA is the same and there are no dental records.

What does DNA stand for?
National Dyslexia Appreciation.

I once took a DNA test, turns out I am 100% going to jail for shoplifting.
I seduced a woman by telling her I could sequence her DNA, but I stopped at the first base when she told me what I would find in her genes.

Not only was Gene Hackman a famous movie star but he also invented DNA decoding.

The Funny Bone

What did the patella say to the femur?
I kneed you.

An Otolaryngologists favourite saying: "The way to a man's stomach is through his oesophagus."

How do you know a spirometer is faulty?
It's expired.

There was a group of biologists who also wanted to start a boy band, they called themselves new kidneys on the block.

A biology graduate was giving a graduation speech, he started with; "I would like to thank my skeleton for all the support it has given me over the years.

What is the most musical part of your body?
The trombone.

One day all the organs in the body had a meeting to decide who was going to be the team leader.

The stomach says, “I should be in charge, I digest the food and give you all energy.”
The heart says, “I should be in charge I am the strongest muscle and I pump all the blood around to give you energy.”
The brain says, “I should be in charge, I am the smartest and give orders to all the body parts.”
The rectum said, “I should be in charge since I remove waste.”
Then all the other organs laughed at him. So, he decided to shut down for a few days and let nothing out. Soon the brain was getting headaches, the stomach was feeling ill and couldn’t digest and the heart lacked the energy to pump blood. All the other organs then decided to make the rectum the leader.
The moral of the story; You don’t need to be the most supportive, smartest, or strongest to be the leader. You just need to be the biggest asshole.

What did the cadaver say to the student?
You stole my heart.
What did the doctor say to her daughter?
The best way to a man’s heart is through his pericardium.

Why did Dr. Anna change her last name to Tommy?
So, she could be called Dr. Anna-Tommy.

What has 13 hearts but no feelings?
A deck of cards.

Two blood cells fell deeply in love, unfortunately it was all in vein.

Did you hear about the doctor that grew vocal cords from stem cells?
The results speak for themselves.

What is the difference between toe jokes and eye jokes?
Toe jokes might be cheesy, but the eye jokes are cornea

A biology teacher asks her class.
"Ok class, what separates the head from the rest of the body?
Little Johnny yells out, "An axe miss"

A biology teacher was teaching students about male anatomy. She was explaining the places and differences in the glands, testicles and epididymis and function of the urethra.
"Miss, I thought that the testicles and the urethra were the same thing?"

"No there is vas deferens between the two."

What does a biologist say when he is giving blood?
B positive.

Randy Researchers

If I was an enzyme, I'd be DNA helicase so I could unzip your genes.

You're so hot, you're denaturing my enzymes.

Want to be my substrate enzyme?

Hey, speaking of biology, let's fuck.

You can be my retrovirus and gag on my pol.

Hey baby, want to form a zygote?

Every time you walk into a room my zygomaticus muscles contract.

I wish I was adenine; then I would get paired with U.

You remind me of amylase, you make everything sweeter.
You breathe oxygen too? Wow, we have so much in common.

Hey babes, I have a smooth endoplasmic reticulum, but you know I want it rough.

I am like a endoplasmic reticulum, I'm rough for your pleasure.

How about I fondle your vesicles while you caress my Golgi body?

You look like you need a little bit of vitamin ME inside of you.

I want to take you home and work on your leucine zipper with my zinc fingers.

You must be the one for me, my selectively permeable membrane just let you through.

Let me be your restriction endonuclease, so I can give you sticky ends.

I am no organ donor, but you can take my heart.

You walked into the room and a sudden protracted cardiac arrhythmia occurred.

Babes, you remind me of cholesterol, you're hurting my heart

Let's unzip your genes and share codes together.

When you're around I go through anaerobic respiration, you just take my breath away.

RNA is my favorite part of my gametes' because it has a U in it.

I'll be like T-cells and protect you.

Babes, you must be a neuron, you have some potential.

You must be a cure for Alzheimer's because you are unforgettable.

Hey, cutie will a little more alcohol catalyse this reaction?

They say the tongue is just a muscle, mine needs a workout so can you sit on my face?

Hey, I need your help to extract some protein from my column?

You give me a jolt of energy greater than my mitochondria.

The inhibitor markers of my DNA must be blocked because I can't stop flirting with you

My hypothalamus must be secreting serotine because baby you're making me happy.

Hey babes, how about I get my ligase to length your Okazaki fragment?

DNA backwards is AND, as in you AND me!!

Babes, be a chromosome with me so we can be a homologous pair.

I don't like to introduce foreign bodies into my body but for you, I'll make an exception.

There are 206 bones in the human body, would you like one more?

I saw an ad in science weekly, it read;

A lonely TACGTAC is looking for their matching congenital pair ATGCATG

Are you a virus?
Because you make me all hot and sweaty.

Are you an amino acid?
The longer we talk, the more my polypeptide chain grows.

Are you a cell?
Because my DNA is all over you.
Are you, my appendix?
Because I want to take you out.

Are you attached to my cardiovascular system? Every time I see you, my pulse quickens.

Do you like science?
Because tonight you are going to sample my DNA.

Did you know your heart stops when you sneeze? Like what my heart does when I see you.

Are you a concentration gradient?
Because I want to go down on you.

Wrong Schooling

Why couldn't the biologist work at a bookstore?
It was a hard cell.

What did the professor say to the biology student when their grade fell from A to F?
You're biodegraded.

Why aren't students allowed in the biologist canteen?
It's staph only.

How many ecologists does it take to change a light bulb?
Four. One to change it, one to write an environmental impact statement and another two to review it.

What did the biology student call a microtome?
A very small, tiny book.

On a university notice board, a question was posed.
'What is 3 X 3'?
The engineer excitedy pulls out his slide rule and shuffles it back and forward and finally determines it is 8.99

The statistician gets his computer out, creates a program, makes a bell curve and comes up with an answer that is in the range of 8.98 to 9.02

The philosopher smiles and writes, "What do you mean 3 X 3?"

The socialist was just happy that everyone was talking about it.

The ecologist turns around and mentions that it is a polyamorous dating ritual.

The medical student pipes up, you know it's 9, right? I know because I memorized it.

So, it turns out the biggest cause of cancer in lab rats is research.

I made a joke in biology class, but no one laughed
I think my thymine was off.

What was the motto of a biology teacher who moon-lighted as a prostitute?
Sex cells.

What university professors are the shadiest?
Biology professors, they have a lot of skeletons in their closets.

What do students who received a fail in biology get on their transcript?
A Bio-D-grade.

Why did the biologist student fail anatomy?
He couldn't cut it.

Why did the biologist student fail the practical?
Because the tutor was sternum.

Little Jimmy failed his biology exam. When asked; "What are inside cells?" His answer, "My Daddy and Uncle Jimmy" wasn't quite right.

Dr. Oades was teaching her biology class at a prestigious high school when she noticed one of the students not paying attention.
"Rose, can you name the part of the human anatomy that under appropriate conditions will grow up to 6 times larger?"
Rose freezes for a second, then slowly a smirk crosses her face.
"Um, Miss? Isn't that an inappropriate thing to ask me? I find that very offensive and I think I will complain to my parents." Rose answered with a smirk.

Unphased by this uppity student's remarks, she turns to another student.

"Lexi, can you name the part of the human anatomy that under appropriate conditions will grow up to 6 times larger?"

"It's the pupil under dim light!" Lexi replied.

"Very well-done Lexi, thank you for that brilliant answer."

She then turns back to Rose.

"I have 3 things to say to you. Firstly, it is quite clear that you haven't been paying attention or doing your homework. Secondly, I think I will have a conversation with your parents about your dirty mind! Lastly Rose, one day, most likely soon, you will be very disappointed."

Little Jimmy went to school and was sitting in anatomy class. The teacher was busy using a poster was pointing out the reproductive parts of the male. Little Jimmy starts to get confused and raised his hand.

"Miss, Miss, that's wrong. My Dad has 2 penises!"

"That's not possible." replied the teacher

"My daddy has a small one to pee with and a long one to brush my mom's teeth with!"

During a biology exam, a question was asked. Explain 3 positives about breast milk.

Little Jimmy answered, contains all the vital nutrients, doesn't need heating and has great packaging.

Little Jimmy is handing in his biology assignment. The teacher starts browsing through it.
"Jimmy where is your appendix?" he asks.
"That is an easy one," Jimmy says as he lifts his stomach and points to his belly.
"Jimmy, what is a specimen?" the teacher askes
"That's an easy one miss, an Italian astronaut."

During a biology lecture, the professor was giving a talk on sex cells and was talking about the high glucose levels found in semen. A new female student puts up her hand and asks, "Excuse me but are you talking about glucose? The same glucose that is found in sugar?"
"Yes, that is correct." answers the professor
"Then how come it doesn't taste sweet?" she blurted out with thinking.
With embarrassment, she packs her things and makes for the door quickly.
"Oh, that is because the taste buds for sweetness are found on the tongue, not at the back of the throat." answers the professor.

Biology teacher: If you don’t stop making puns about plants, I will send you outside.
Student:Don’t worry I’m Leafing?

I learnt today that the biggest organ is the human skin. I always thought it was in the town cathedral.

“Me biology teacher told me all ants are females!” said little Jimmy to his mum.
“Oh is that so?” asked his mum with a curious look on her face.
“Yeah, the males are called uncles!” Jimmy continues.

The biology teacher handed back the exams on insects. Little Jimmy had received an ‘F’.
“Are you chitin me?” he says to the teacher.

My wife and I decided to make a pact, that we would make a list of all the people who could sleep with if we got a chance.
She chose Chris Hemsworth, Channing Tatum, Pierce Brosnan, and Zac Efron. I chose her sister, her cousin that was at our wedding, the local bar girl, and our son’s biology teacher.

In biology class I learned the dangers of smoking cigarettes today, thank God I switched to crack last week.

How did the frogs feel about the biology class?
They were pithed

A primary teacher was teaching the class biology. She was saying to her class that humans are the only animal to stutter.
Little Jimmy puts up her hand, "Miss, miss that's not true. My neighbour's cat stuttered once."
"I am sorry Jimmy, that's not possible."
"Yeah miss, it's true. It jumped into our yard and our Rottweiler saw it. Then the cat started going fffft fffft fffft and before the cat could tell it to fuck off, the dog ate him."

A biology teacher and a mathematician get together and decide to make a plan to save the planet. They called it the Al-gor-ithm.

A biology teacher was asked to step in for a history teacher, and she did very well. The first thing Miss Kuhn did was get the students to draw a picture of Abraham Lincoln and then label the parts.

High school was exactly like Darwin's theory, I wasn't selected for anything.

A student was up to his last question in his botany exam, but he couldn't finish. He was stumped.

I would have gotten an 'A' in all my exams if it wasn't for genetics, now I am going to have to go home and tell them I am stupid.

At a university, who is in charge of the finances of the biology department?
A buy-al-agist.

There was a senile old biology professor who would often get distracted and just start telling dirty jokes. There were a number of students who got sick of this behaviour and decided to walk out the next time he started.
The following lecture he started, "That reminds me, did you know there is currently a shortage of prostitutes in Holland?"
The group of disgruntled students closed their books and started walking out.

The professor called out to them, “Ladies and Gents, no need to rush, the next plane leaves at 7 in the morning.”

I remember the time I brought home a report card with an A+ on it. When my parents looked at it, all they said was, “Who the hell is George, and you don’t even do biology!!!!”

A biology teacher asked her class, “What is relative dating?”
“Isn’t that when rednecks kiss their cousins?” replied one student.

A quite talkative student was asking the biology teacher, several questions to try and get a rise out of them
Student: So, if a snake bites me and I die, would that be called venomous?
Teacher: Yes, Johnny the snake is venomous
Student: What if I bite the snake and it dies?
Teacher: Then you’re venomous
Student: What if the snake bites me and someone else dies?
Teacher: That is correlation and causation
Student: What if the snake bites itself and I die?
Teacher: That is called voodoo

Student: What if the snake and I bite each other and neither of us dies?
Teacher: Then you're into weird kinky shit and probably going to hell.

A student was up all night studying for his zoology exam, looking at all rare endangered species around the globe and their locations. It was the last exam he had to do before he graduated. The next day, being summer he walked into the auditorium ready for his exam with 300 hundred other students wearing his lucky shorts. He looked upon the stage and saw 8 platforms. On each platform, there was a bird with a sheet over it. You could only see the bottom of their legs and feet.
The professor gets on stage and explains. The students only get to look at that part of the animal. They then must describe the function of the feet; from there they must deduce the species and family names. The student looked complexed. He started sweating he had no idea. He started thinking of the injustice, it wasn't fair he had studied hard but did not have a clue. He started getting angrier and angrier until he reached a breaking point. Standing up he threw his books on the ground and yelled, "FUCK THIS!"
The professor was in a state of shock as the student was walking away.

"What's your name?!" he yelled after the student.
"Well look at my fucken legs tell me and what family I come from!!"

Why does getting a B for your biology practical make the class difficult?
Because a frog is easier to dissect.

Stereotypes

Biologists, thoughtfully staring down microscopes for no apparent reason since the 1600s.

I have a degree in biology, it's BS.

Why would you study biology? It's not like you have a life.

Where do biologists go to church?
A cysteine chapel.

What do you call a group of biologists at church?
Biomass.

What did the biologist's name their twins?
Jessica and Control.

My biology partner was dissecting some insects. I had to tell them that their fly was open.

A biologist, a physicist and a chemist go to the ocean for a field trip for the first time.

The biologist is overcome with this new environment and wants to study the seaweed fields. Excitedly, he runs straight into the ocean where the undercurrent grips him and takes him out to sea. He drowns.
The physicist ends up in a hypnotic state, just starring at the waves. He is lost in thought looking at the fluid dynamics and wave motion. Counting to himself, he slowly walks into the ocean where the undercurrent takes and drowns him as well.
The chemist takes out his notebook and writes:
Biologists and Physicists are both water-soluble.

Chemists tell you that you are made from 60% Oxygen, biologists say that you're made of 70% water and Physicists say you are 99.99% Space, but I'll say you are 100% cute.

A biologist and a physicist fell in love and got married but get a divorce very early. Turns out the chemistry wasn't right.

My attempts at a teaching career:
I started teaching history but that's in the past.
I tried teaching chemistry but there were elements I couldn't understand.
I tried teaching biology, but my heart wasn't in.

I didn't like teaching maths; something wasn't adding up
I couldn't teach German, I was the wurst.
I am now trying physics, it seems to have potential.

You and I have some chemistry, do you want to do some biology

What certificate do real estate agents' study?
Homology.

During a work meeting, one of the staff yells out, "Three to five cellular layers of skin."
"What?", says the boss sharply
"I am just saying what is on everyone's lips."

One day George is sitting in the laboratory talking to a work colleague.
"This tennis elbow is killing me," he says, "I am going to have to see a doctor."
"Oh, don't do that, that is way too expensive, the pharmacist has this brand-new machine. All you do is drop a urine sample in, it then analyses it and tells you what's wrong and how to fix it. It will cost about $10. Super cheap."

That afternoon George goes to the pharmacy, goes up to the machine. He inserts the cup in the square opening, slides the glass door, inserts $10 and waits. The cup gets lifted into the machine and the lights start buzzing. After about 5 mins a ticket shoots out, and it says;
You have tennis elbow; you need to not lift it or put any unnecessary strain on it. Soak in warm water every night and take anti-inflammation tablets for 2 weeks.
George is so amazed at the machine. He decides to put the machine to a real test for fun. He grabs a cup, puts water in it and then gets urine samples from his dog, wife and daughter and then for good measure he also masturbates in the cup. He goes back the next day and tests out the machine.
When the machine finishes its testing, it spits out the results which read as follows

1. *The tap water does not have enough fluoride in it. Add tablets to the family's diet*
2. *Your dog has ringworm, use the anti-fungal medication*
3. *Your wife is pregnant with your neighbour's seed, get a lawyer*
4. *Your daughter is using ecstasy, get her into rehab*
5. *Finally, if you keep masturbating your tennis elbow will never clear up*

A biologist had triplets, they were called Poly, Ethel and Ian.

How does a biologist pass a test on excretion?
By a process of elimination.

What kind of anti-biotic do you give to an Italian?
Amoxsicillian.

A biologist, a physicist and a mathematician are at a café together having brunch. They notice two people walk into a building across the road and then 3 people leaving. The physicist remarks, "We must have made an error in our original estimation."
The biologist says, "No, they must have reproduced."
The mathematician remarks, "Now we are at -1, if someone enters then the house will be empty."

A physicist, a zoologist and a statistician go hunting in the forest when they come across a deer. The physicist takes a shot and missed exactly 2 ft to the left of the deer. The biologist then takes aim and missus exactly 2ft to the right of the deer. The statistician then yells, "Eureka we got it!"

If you wake up in a science lecture, how do you know which class you're in?

- If it moves it's biology
- If it stinks it's chemistry
- If it doesn't work its physics

A guy jumps out of the alley and pulls out a gun on an unsuspecting couple.
"Give me all your money or your biology" he growls.
"Don't you mean history?"
"Don't change the subject!"

How do you tell the difference between someone who is a biologist or someone who loves BDSM?
Ask them what dominant means.

The Mars Rover suddenly stopped transmitting signals back to earth, so NASA decide to bring back the Mars Rover. It cost millions and was a really difficult task. After well over a year, it was finally back home on planet Earth, and all the scientists were eagerly gathered around it, patiently waiting to turn it back on to reveal the missing data. It exploded and sent debris everywhere. Luckily there was only one casualty, Sylvester, the office feline and employed rat catcher.

Moral of the Story: Curiosity killed the cat

A botanist, an ecologist and a senior scientist were scouring the hot jungles of Papa New Guinea looking for rare or new species of either wildlife or plants. They mainly discovered dysentery, sweat rashes, box flies and the always ever-present humidity that clung onto their frames. Then one day they came across ruins, the ruins seemed to look like something out of an Aladdin or Sinbad movie, Middle Eastern and out of place in the lush humid rainforests of Papa New Guinea. They moved hesitantly inside and into the middle of a building, there was a pedestal with a lamp.

When they approached, one of them tripped over a tree root causing all 3 of them to stumble and fall onto the lamp. Just as they touch it, WOOSH, a genie appeared from it.

"Since all three of you touched the lamp at the same time, I will grant each person one wish." boomed the Genie.

"I want to be on a deck chair next to the Caribbean Sea, getting a massage and someone serving me cocktails," piped in the botanist.

Then woosh he was gone.

"I want to be in a spa, with two beautiful companions on a veranda in the Swiss alps," yelled the excited ecologist.

Then woosh he was gone. The senior scientist looked around at the out of place structure and couldn't help but get more and more curious. Why would such a structure be located here?

"Ok I'm ready for my wish," said the senior scientist thoughtfully, "after lunch, I want the other two back here."

Moral of the story: In important meetings always let the boss speak first.

What did Gregor Mendel say when he understood genetics?

Woo-pea.

Why did the chicken cross the road?

Darwin: "Well only the fittest chickens made it across, and it was the next logical step after they came down from the trees."

What did the other monks say to Gregor Mendel?

Peas for dinner again?

What prize did Gregor Mendel receive?

The noble peas price.

A man is standing on the edge of a cliff looking down. His wife walks up and joins him.

He says to her, "I bet if we jump off together, I will be the first one to hit the bottom."
"I doubt that," she answers.
So they both jump off at the same time to see. Who wins?
Charles Darwin

The Realistic Glossary

A

abiogenesis – someone who is good at biology
active transport – a bus that works
adaptation – when it is not the child's natural parents
adenosine triphosphate – what cells use for boom energy
advection – an advertisement for arrows
aerobic – a sport which involves yoga clothes
agglutination – when lonely cells come together
algae – technically your great great great ……………………. grandmother
allele – French for Ally
alveolar – when Ally lies in French
alveolus – when Ally is horny in French
amphibious – Richard Branson has the record for crossing the channel in one of these vehicles
amphiphilic – the ultimate bisexual molecule
Anaxagoras – the first Scientologists
Anaximander – he got confused with evolving from fish but then magically people just spawned from nowhere.
anther – married to me uncle
antibody- the Spartans of your immune system
antiport – otherwise known as not a port or an airport

apex predator – Jeffery Epstein
archaea – like bacteria, but not
artificial selection – dog breeding
astrobiology –space animals
Avery, Oswald – one of the discoveries of DNA and bane to all murders the world over.

B

bacteria – found everywhere and apparently not as cool as archaea
bacteriophage – when a virus enters a bacterium and then burst out. Think of the movie *Alien* but on a tiny, small scale
Bateson, William – coined the term genetics and is famous for being Mendel's publicist.
binary fission – the boring way to reproduce
binomial – attracted to both nom and mal
biogeochemical – something that couldn't make up its mind in which branch of science it belonged to
biomass – measuring your mama's butt
bioregion – a where a person's yeast infection is found
boundary – when you say no
breeding – the fun way of reproduction
breeding population – a swinger's club

Brown, Robert – watching pollen he discovered physics that only Einstein could prove

C

carbon -14- dating – the legal age before carbon goes on a date

centrioles – small sausage rolls

Chambers, Robert – the guy who no one remembers, just before Charles Darwin

Chargaff Ewin – discovered that DNA runs on time (thymine)

Chase, Martha – discovered some stuff then forgot she did

chemiosmosis – osmosis cells undergoing chemo

chloroplast – plant cells solar panels

chromatin – when crows have sex

chromosome – when more than 2 crows are having sex

circulatory system – the heart sharing the love around

classification – animal discrimination

concentration – what a physicist does when they try to understand a biologist textbook

concentration gradient – when a physicist draws a graph to understand the biology textbook

conservation biology – an area where endangered biologists are placed, and their breeding habits noted

Cope, Edward Drinker – wrote over 1200 papers, hey if you are writing that much you are bound to get something right

Correns, Carl – German botanist and geneticist, he could have possibly discovered the secret to the super solider serum but most of his work was destroyed in allied bombing

cortisol – what bodybuilders try to suppress

Crick Francis – One of the discovers of DNA and a boon to patiently suits across the world

CRISPR – forgetting what's in the oven

cytokinesis – when one becomes two

cytoplasm – cell goo

cytosol – more gunky cell goo

D

de Vries, Hugo – another one of Mendel's publicists, meaning he re-read the work and republished it.

decay – next to the de bay

descent – something that smells bad

diffusion – combing the French food with foods from other cultures

DNA – Don't Need Anal

DNA helicase – loves to unzip things

DNA Ligase – zips things back up

DNA Polymerase – an enzyme that has is married to more than one catalyst
Dominant – Dom's auntie
Dominant – when a biologist gets kinky

E

effusion – farting but saying it with science
electrolyte – should not be used for watering crops
endemic – native to a local environment, for example, my ex was a succubus straight from hell
endocytosis – when a cell eats, think of the movie *The Blob*
endosymbiosis – when an organism lives entirely within another, just like a foetus
environment – setting the ambience before date night
enzyme – are things biologists invented to explain harder things.
epidemiology – the study of epic spiders
eugenics – the Nazi party was a big fan of this
eukaryote – when the brain of a cell has a skin
evolution – when superior mutants become the norm
excitable – when you hear the ice cream truck music
extinction – dinosaurs didn't drink alcohol and look what happened to them

extrapolate – the ability to get any data and change it to fit your narrative

F

facilitated diffusion – when your partner drags you off against your will to that fancy restaurant combining French foods with other blends, and all you wanted was a burger

filial – science naming things to make it difficult, offspring, is that really that hard of a word?

fixity of species – opposite of evolution sadly the idea of X-men is not even remotely possible with this theory

fluorescence – cheap lighting

folk taxonomy – naming the village people

fossil record – your grandparents birth certificate

Franklin Rosalind – the woman who managed to discover the structure of DNA but was denied a Nobel Peace Prize. Twice.

G

gene –a person who played in the band called Kiss

gene drive – when Gene drives and talks in the 1st person at the same time

genetic drift – why there will be more and more attractive people in the future

genome – Gene Simmon's garden gnome

genotype – the women Gene goes for, which is all of them

genus – Gene's kids

glucose – what goes straight to my thighs

Golgi apparatus – acts like a postal processing plant but for proteins in cells

Goodall, Jane – had a weird relationship with chimps, also discovered they had entered the stone age.

Gould, Stephan Jay – massive supporter of evolution

Graham Thomas – made a law about how farts diffuse in air and named the law after himself

Griffith Frederick – his bacteria experiments left more questions than answers

H

Haldane, J.B.S – actually, I really don't want to make a joke about this guy, he's a genius, he was the next step in evolution, he put Darwin's theory and Mendel's genetics together and in his spare time worked out how to cure compression sickness for divers, and also noticed someone had a beetle fetish

haplotype – genes we inherit from the 1 parent, usually the happier parent

Hardy Weinberg Equilibrium – if there is no evolution then the number of Wieners stay the same

hemagglutination – when red blood cells hug

heredity – when you find out you are the postman's kid

Hershey, Alfred – got a Nobel Peace Prize for using blenders.

heterozygous – as this little man in a Balinese t-shirt shop told me about his Versace handbags, "same, same, but different."

homeothermic – when two men who love each other buy a thermos flask

homozygous – look at Heterozygous, but the saying is "different but same, same."

Hooke, Robert - while chasing Peter Pan using explosives, Hooke came up with the equation of elasticity horticulture – you can make a lot of money especially if you set something set up in your roof

Human Genome Project – first look at how to make human DNA experiments

Huxley, Thomas Henry – Darwin's bulldog

hybrid – a car that runs on fuel and electricity.

hydrogeneration – what happens when you eat that extra chocolate bar, you get fat

hydrophilic – a fetish for water

hydrophobic – scared of water or dams

I

immunity – can be natural or injected, when it is injected, it usually has had at least a minimum of 3 long term safety studies to go along with it

in vitro – in a test tube, what do you call a test tube babes worst nightmare? A dingo with a straw

inherent – a lesbian affair between a niece and aunty, (in her aunt).

inheritance – what you get when someone dies

inherited – baldness, need I say more?

invertebrate – something lacking backbones, like our politicians

K

karyotype – taking pics of chromosomes

Knight, Thomas Andrew – a mad gardener

Kerbs, Hans Adolf – another one of the scientists that Hitler chased away

L

Lamarck, Jean-Baptise – Darwin's opposition and proved wrong

Landsteiner Karl – part scientist part vampire

leukocytes – white blood cells that give a new definition to bear hugs

Levene, Phoebus – thought that DNA was much more complicated than it was, trying to prove this he ultimately discovered a lot of things about DNA but also proved himself wrong

lichen – werewolves

limiting factor – money

limiting nutrient – running out of alcohol on boxing day

lineage – only important if you're royalty

Linnaeus, Carolus – invented the classifying of organisms, was also a doctor to the Swedish royal family, i.e. lineage *see the previous statement* lipids – fat, like the salted fat on lambs chops, mmmm so tasty

M

Macleod, Colin – apparently another biologist that did something with DNA and got a Nobel Peace Prize. Also looked at another major murderer of soldiers during WW2, typhus fever, malaria, and pneumonia.

macronutrient – big food like a 72oz steak

McCarty, Macyln – good friend of Macleod and even got a Nobel Peace Prize. Also, part vampire and obsessed with proteins in the blood

mean free path – to become Zen

Medawar, Peter – a pioneer in organ transplants and skin grafting, because of this man's success every time my mother-in-law kisses me on the cheek, she is really kissing my ass

melanin – natural sunscreen, also caused a sunscreen ad to be cancelled in the mid-'90s for being racist

membrane – where me thinks

membrane pump – me brain's pump, maybe me heart?

Mendel, Gregor – had a healthy diet of peas, he ate so much he knew the shapes and sizes quite intimately and even knew what the peas offspring would look like metabolism – what lazy people try to blame for being fat

micronutrient – a small feed that leaves you hungry. e.g., celery

microtubule – having a very tiny dick

Miescher, J. Friedrich – discovered concentrations of CO_2 affects the rate of breathing. Used the excuse of not becoming a doctor because he was half deaf but being deaf never stopped Beethoven

mitochondria – the engines of the cell

Morgan, Thomas Hunt – another biologist who got famous by publishing stuff from Mendel

mutant – someone with superhuman powers

mutate – when a hero with superhuman powers starts to lose control of their humanity

N

Nageli, Carl – Swiss botanist who wrote about inherited characteristics and the legend didn't mention Mendel once in his book

natural selection – this is aided by many things, the plank challenge, taking selfies on cliffs and 40km/hr speed zones outside high schools.

neanderthal – humans from a much simpler time, but a bitch of a name if you suffer from dysarthria

net movement – towing a net behind a boat

neutrophil – determines how brave you are, how strong your backbone is

Nightingale, Florence – started the Nightingale School of Nursing which teaches caregiving principles, what a novel idea, caring for your fellow man

nomenclature – naming stuff, that's it! Naming stuff, nice and simple

nucleic acid – an acid that comes from nuclear reactors, might either kill you or give you superpowers

nutrient – if you're a POW then a cockroach is one

nutrient influx – Christmas lunch

O

Okazaki fragment – a bit of Japanese DNA

organelle – cell organs

organism – making biology teachers accidentally say

orgasm since 1910

osteological – the funny bone is not one

outbreak – A 1995 movie starring Dustin Hoffman

P

palaeobiological – what happens when a biologist and a paleologist have child panspermia – Ancient Greek theory that all the seeds of plants rained down on us from asteroids, this theory is starting to make a comeback

parasite – and organism that lives off another organism. Examples include malaria, ticks, welfare bludgers, politicians, and my ex-girlfriend. Parasites even had a song written about them by Offspring, '*Why Don't You Get Job.*'

Pasteur, Louis – a microbiologist who found out bacteria can make people sick and found out that boiling things killed bacteria. So now boiling things will forever be known as pasteurization.

Pauling, Linus – had shares in Vitamin C

peptide – a short tide

phenotype – what something looks like

phospholipid – fat that can glow in the dark

photosynthesis – the first solar power cells

phylogeny – evolutionary tree, turns out it isn't a tree at all.

polar – north and south have one each

pollen – responsible for hay fever, when you think about it, plant sperm is making your nose sneeze

pollinate – what pollen tries to do to your nose

population – in humans this always seems to increase 9 months after a power outage.

preening – what teachers should never do

progeny – a Finnish heavy metal band

protein – what bodybuilders can't get enough of

R

random walk – what happens when you're trying to walk home drunk from the bar

Ray, John – looked at plant babies and named them depending on if they had 1 or 2 leaves

recessive – the other half of a kinky biologist couple

Remak, Robert – uing a microscope found how cells multiplied from existing cells, also denied a professorship because he was Jewish. That's right he also lived in Germany

replication – copying someone's work or what science calls peer review.

reproduction isolation – masturbating when you are alone

respiration – breathing but complicated

resting membrane potential – when the membrane is relaxing, catching some R & R

rookery- a breeding place for birds, also where people met if they have a bird fetish

rough endoplasmic reticulum – has the word rectum in it.

S

Saint Hillarie, Isidore Geoffroy – loved looking at animals in their natural habitat, but also had a fetish for congenital abnormalities in animals

sanitation – having a healthy septic system, why was it such a difficult concept to understand that you shouldn't shit where you eat?

saponification – turning fat lady's ass into soap and selling it back to them

Schwann, Theodor – loved beer and bread so much he studied yeast and showed plants and animals are made up of cells

serology – just because someone is giving a doubleheader in the toilet doesn't make them a serologist

Spallanzani, Lazzaro – took lots of 'herbal' tea and found spontaneous generation in broth and herbal infusions, studied the circulatory system, ~~blamed~~ pioneered artificial insemination, and found out that bats navigated via echolocation

species – which animals can have sex and still produce fertile offspring.

sugar – the real reason why people get fat

symbiosis – when 2 creatures live together and support each other mutually, unlike my ex

T

taxonomy – trying to not pay tax

Theophrastus – Greek philosopher who studied plants and loved living on the island of Lesbos

trait – a certain characteristic an individual had, for example, Bernie Madoff's loved conning people out of money. Traits aren't necessarily nice.

Tyson Edward – tried to cross-breed monkeys and apes and compare them to people, the original *Planet of The Apes f*an.

U

uropygial gland – the glands birds use to pee on themselves

V

vaccine – you know it works when it stops the spread of disease and stops thousands of people dying from the disease and thousands of people don't die when they have taken it.

vein – what junkies can find in 2 secs but what my nurse takes 50 goes trying to hit one

Venter, Craig – developed techniques that allowed rapid sequencing possible. Later founded Celera Genomics which would help garden gnomes with their economics

vertebrate – a backbone, something a lot of people do not have any more

vesicle – what Vin Diesel nicknamed his testicles

villi – I will never own one in the Caribbean unfortunately

virus – this is 2022, the beginning of year 3 of the 15-day of flattening the curve.

viscosity – how sticky or runny something is. For example, the viscosity of semen is 9.35 centipoise

W

Wallace, Alfred Russel – was a fan of Darwin theories and tried to follow in his footsteps. He got 'lost' in the Indonesian Archipelago for 8 years

water-insoluble- it doesn't melt unlike the wicked witch of the west; she was very water-soluble.

wean – to get off breast milk, but men can never seem to get over boobies

Wilkins, Maurice – worked with radar and the Manhattan Project, later turned his attention to DNA, he was the original Bruce Banner but unfortunately, no accidents happened, so we never got the incredible hulk.

Y

yeast – this little beautiful thing has been getting the human race drunk for thousands of years.

Z

zygote – a goat that came from Zion, or the offspring of when a farmer fucks his goat.

Please don't forget to visit www.unconventionalpublishing.com.au and to receive your free adults pnly joke book.

Authors Message

Hi, there.
Thank you for taking the time to read this book. Coming from a science background I have been wanting to write these joke books for a long time and unleash the inner nerd in me. I hope you had just as many laughs as I did in writing this book. If you did enjoy reading this, please click on the link to leave a review from where you bought this book, that would be greatly appreciated.

You will also find another joke book on the website, one that is not for sale, one that is not for the fainted heart, one that can be considered a bit risqué and politically incorrect. Consider this a gift for taking the time to purchase and read my book but be warned, only download it if you are not easily offended.

www.unconventionalpublishing.com.au

It was mentioned in the introduction that this book will be part of a series that will be coming out. If Science jokes tickle your fancy and you haven't had a chance to read the first book, look at *The Fantastic Book of Chemistry Jokes.* There will also be a clothing line with some of the one-liners on them. If you want to see any other professions being roasted or want a particular joke on a shirt, please let us visit our website and let us know.

Kind Regards
Shane Van

Printed in Dunstable, United Kingdom

a87a4e26-9a16-4316-b527-826c23f1f994R01